The Use of
Stored Forages
with Stocker and
Grass-Finished Cattle
by
Anibal Pordomingo

A division of Mississippi Valley Publishing Corporation
Ridgeland, Mississippi

First printing March 2016

Library of Congress Cataloging-in-Publication Data

Names: Pordomingo, Anibal, 1963- author.
Title: The use of stored forages with stocker and grass-finished cattle / by Anibal Pordomingo.
Description: First edition. | Ridgeland, Mississippi : Green Park Press, [2016] Includes index.
Identifiers: LCCN 2016001294 | ISBN 9780986014727 (alk. paper)
Subjects: LCSH: Beef cattle--Feeding and feeds--Argentina. | Hay as feed--Argentina. | Forage.
Classification: LCC SF203 .P67 2016 | DDC 636.2/130982--dc23
LC record available at http://lccn.loc.gov/2016001294

Cover design by Steve Erickson, Madison, MS
Manufactured in the United States of America

TABLE OF CONTENTS

A note from the editor...

It might seem strange that a publication that has been preaching a no-hay approach to livestock raising would publish a tract on the positive benefits of hay feeding. Of course, the difference is we are talking about supplementing pastures with hay and not replacing them and we are talking about finishing cattle.

This is very different from cow-calf production. Finishing cattle to the High Select/Low Choice grade on forages alone is not natural and requires unnaturally good forages and management. The high rates of gain required to produce marbled beef often require supplemental dry matter and supplemental quantity.

Due to the high quality of stored forages needed with finishing cattle, it may be wise to buy forage-tested hays rather than try to make them yourself. This is particularly true if you live in a humid area of the country. Keep in mind that finishing animals require the same forage quality as a dairy cow, so dairy quality hay is what you should look for and buy.

This tract was written by Anibal Pordomingo, a rancher/researcher from Argentina, who has taught our Stockman Grass Farmer school on grass finishing for many years. We lead off with a tour of his farm in the central Pampas region, which is very similar climatically to central Oklahoma. Note that even with his nearly year around hay supplementation, the total amount fed during the entire stocker/finishing phase is only equal to one roundbale.

Allan Nation
Stockman Grass Farmer

5

A look at Anibal's family farm

My brother, sister and I run a 450 acre beef grass finishing operation in the Argentine pampas. We do not share ownership with others and now own all of our land freehold. Even our dad runs his own cow-calf separate from us in a small property close to town. The three of us have off-farm day jobs and that creates restrictions full-time ranchers do not have.

My sister and I work at an agricultural research station from 7:00 to 4:00, and my brother works in his accounting and teaching business. This means we must run our own operation in spare time during the week and on the weekends.

Unlike many of our neighbors, we do not include grain cropping in our farm. We do not like the high financial risk of grain farming and its marginal economics. We like the cattle business and enjoy doing it.

We run an extensive low-input program. We use no fertilizers or pesticides, and alfalfa provides all of the nitrogen for our forage chain. Our basic business plan is to produce well finished beef using a low-cost low-investment model and to put all our resources into cattle growth. Therefore our investments are limited to pastures, cattle and time. This last resource - time - is for us the most limiting and we try to use it as efficiently as possible.

Over the years, we have found we cannot directly exchange time for equipment or infrastructure. We found that owning "iron" (farm equipment) actually demanded more time and likely has a greater opportunity cost than investment return at our farm. Therefore, we own only a 40-year-old tractor and a disk plow. Water is all windmill pumped to above-ground tanks and then piped to water troughs in the fields and paddocks.

We contract out all the haymaking, most of the land disk plowing and planting work. We set up the temporary fences, read the pasture availability and rotate the grazing, deal with the animals' health and do any doctoring if needed. We use one-wire, high-tensile, gate-less electrified fences whereby the wire is held up by a temporary electrified pole (with insulated handles and botton) to allow cattle and vehicle access the paddock. The pole is called a *vela*. We usually layout large semi-permanent electrified paddocks that are further subdivided with temporary poly-wire electric fences.

Vela pole with insulated handle and wire at the top.

As much as possible, we want to be the only two people around the cattle from arrival to loading for sale. Even the

Bars on bumper for driving over electric wire fence.

vehicles are always our own and nobody else drives around
the pastures or the cattle. We have found that this strategy
minimizes stress on the animals. We do not do any roping or
rough handling on the field. Any individual handling gets done
in the corrals and by just the two of us on foot.

We only do the obligatory vaccinations (hoof and
mouth, 3-way) and some specific animal treatment for injuries
or particular infections, pink eye, which we have had only two
to deal with in an average year. We buy the stocker calves from
a cow-calf operation in our family, and concentrate on finishing
stockers.

We currently run about 250 steers year around. The
growing-finishing period for steers ranges from 12 to 14
months and 8 to 9 months for heifers, which means that steers
are harvested in the range of 18 to 20 months of age and heifers
in the range of 14 to 16 months of age.

We have a simple forage chain:

1) Alfalfa pasture from spring to fall.

2) Winter annuals (planted in early fall) for winter and early spring.

3) Greenleaf corn for grazing in summer.

While we try to feed as little hay as possible, we will supplement with alfalfa hay whenever forage is short to keep our gains from falling below our target of 1.3 lbs per day. This supplemental feeding could occur at any time of the year. We try to produce as much alfalfa hay in the summer as possible. Our hay target is one round bale (1200 lb/bale) per head.

We try to bring stockers to the place with a priority on an arrival date that matches available forage more than size or age of calves. Six to 10-month-old stocker calves fit the program well. Common year's average daily gain for all steers and heifers sold is 1.60 to 1.70 lb/day with 1.8 to 2.0 lb/day during the last two months of the finishing period.

We sell finished heifers at 780 to 800 lb/head (57.5 to 57.8% hot carcass yield) and steers at 1100 lb/head (58.5% yield). At the cow-calf operation, calves are generated from Angus cows sired by Hereford or Shorthorn bulls of fairly predictable performance. We select the original cows and bulls at the same places that provide the calves to us. We believe we have created a biotype of steer and heifer that suits our program and market well.

Our experience has been that the crossbred calves tend to outrun in growth rate, yield and finishing the straight-bred Angus by 20 to 22 percent. We have measured this difference at the farm in simultaneous comparisons. Picking the proper bulls for the cows is a critical aspect of successful grass finishing and care should be taken to select for a grass-friendly biotype.

Monthly Real-time Data

We do a real-time economic and cash-flow study monthly. And, every month we chart out the probable worst case scenario (weather and price shifts) and the strategies we would take to ameliorate it. Paperwork seemed to be over kill at first, but the exercise has been a real help to control the process and even our excitement and depression.

We now know where and with what efficiency every dollar was expended and the real origin of returns. For example, was our profit or loss coming from production or because of our inventory appreciation or depreciation? This is important to know because production we may be able to control, but appreciation or depreciation of assets depends on many economy factors we cannot control.

We sell grass-finished beeves all year long, with some concentration in late winter and spring as finishing cattle on winter annuals is the easiest program. Late fall is usually the time of the year when we sell very little as it is a very difficult time of the year to grass-finish cattle.

We have well-kept working corrals and permanent fences and a well maintained permanent electric fence net. Our farm does not have much of a headquarters or facilities. We live in town, so there is no fancy landscaped park or a large house. If it benefitted the business we would invest in better headquarters, but so far we find that what we have is about what we need.

More than anything, the overall strategy is based on land use to minimize inputs. Land rotation implies creating a mosaic of pastures that sequence summer and winter annual ones which rotate with perennial alfalfa based pastures. It is central to incorporate symbiotic nitrogen with legumes and soil structure with grasses. Hay is part of the system but in a minimum amount to maintain stability. It is fed on pastures to redistribute nutrients and avoid nutrient export.

Considerations for finishing on stored forages

Practical experience indicates that chances for producing marbled beef off pasture with stored forages will be greater if the animals have been well fed on a high quality, green pasture before going into a full stored forage diet. Being that marbling is a fairly linear process (under a good plane of nutrition), with good green season gains we do not need to rely on the last 60 days to lay on the animal sufficient intramuscular fat. A quick and intense final finish is just too much to expect from stored forages and energy-limited by-products. Therefore, the success of finishing on stored feeds will always be greatly determined by the animal's performance during the previous phase on pasture. *CONTINUOUS gain*

It is best if a finishing program does not have to rely 100 percent on stored forages beyond two months time. Having to finish on stored forages for a longer period requires a prodigious fine-tuning of animal genetics, age and diet. For most programs, it will be far easier to carry the animals as stockers on the stored feeds, and then finish on subsequent winter annuals and spring pastures.

Genetics and age play a central role in the success of this program. High-framed animals will be more difficult to finish well even under excellent circumstances. In winter restrictive environments, we may want to plan on finishing the low frame, small animals during the first warm season after weaning, but we should not expect to finish the larger ones during that first fall or winter. It will be far easier and more economical to do it on the second spring or summer, at 24 to 28 months of age. This concept does not imply a purposely-restricted feeding program, just a less demanding winter program that can increase the chances of a longer green grazing period before processing.

Contractor making hay.

Finishing Light or Heavy Affects Beef Quality

Research has pointed out the advantage of harvesting cattle at young ages to guarantee tenderness of all cuts. But, research has also shown that the age for reliable tenderness of grass-finished animals could be well extended to 30 months of age (at slaughter) if the beeve is not exposed to painful feed restrictions (actual weight loss) during its early life, or to long periods of restriction during the stocker phase.

Young animals, harvested lean, will have no marbling and be too thin in backfat. Too little backfat leaves part of the carcass uncovered and unprotected from post-harvest fast chilling issues, meat water loss, and mold growth and the consequent development of undesirable flavors (gamey taste) during aging. With a well-fattened animal this aging mold is trimmed off and discarded and the underlying meat's flavor is unaffected by mold. Without the fat, the mold grows directly on the meat and the meat flavor is greatly affected. This off-flavoring is probably the biggest problem the North American grass-fed beef industry faces and it is one that should be corrected to avoid bad reputation.

It would be far better to grow such lean animals for extra six months (if needed) and harvest them well finished, rather than harvest the animal young, but extra lean. Alternatively, a good strategy for extra lean meat is to shorten the aging period. This would help to retain moisture and would produce fewer off flavors. Alternatively, extra lean, short aged meat can be ground into hamburger meat.

As progress in easy-finishing beef genetics is made in North America, animals will become much more flexible and adaptable to the system. Easy finishing requires early maturing and easy marbling cattle. Once we have such animals, the current heated discussions on flavors, tenderness and fatness should become irrelevant.

Are Fibrous Feeds Enough to Create Omega-3 and CLA-Rich Beef?

Substrates in the diet are needed to build fats rich in omega-3, CLA, and antioxidants in beef. Fiber (forage) creates a proper rumen (digestive) environment for synthesis of precursors, but substrates in the forage increase the supply to the lower gut. Green growing forages are the best synthesis promoters and substrate suppliers. The greater the proportion of green leaves in the animal's diet, the larger the chances of improved concentrations of desirable functional compounds in fat (omega-3 lipids, CLA, fat-soluble vitamins). This is even more so if cool-season grasses, legumes or cool-season forbs are included in high proportion.

Unfortunately, stored forages (ensiled feeds mainly) lose these precursors and therefore result in lower concentrations of omega-3 lipids, CLA and fat-soluble vitamins in beef. Silages are generated in a high reducing environment (lack of oxygen). Weak acids in silages help the breakdown of fibers but saturate all fatty acids and unsaturated metabolites. The rumen environment saturates fats to a large extent, but the process is more complete if the feed has an acid treatment (ensiling) to start with. Mixtures of silage with quality hay or pasture could help to reduce the saturation effect.

There has been much discussion in the upper Midwest

on the suitability of feeding dried distillers grains (DDGs) in a grassfed program. The argument is that the processed grain has lost its starch component and therefore will not affect the fatty acid content or CLA content of the meat. So far, no significant research has been reported on the effects of DDGs and similar ethanol by-products as supplements on stored forages or green pastures. However, it could be hypothesized that the response will be related to the rumen availability of precursors and the effects of such by-products on rumen pH and fiber fermentation. Differences in lipid contribution between corn and by-products of corn would not be expected. Therefore, lipid profiles of intramuscular fat fed DDGs for an extended period in the absence of green pasture would probably start to resemble those of grainfed animals.

In conclusion, when fed solely as supplements and not as the entire feed supply, stored forages will not be contributors of precursors but will not negatively affect the response to green pasture. Quality and intake of green pasture under such conditions would likely dictate the response in performance and omega-3 lipid profile. However, when fed in the absence of green pasture for extended periods, animals finished on stored forages or DDGs will probably have significantly less omega-3, CLA and fat-soluble vitamins.

Grazing hay

Hay is the most common tool to transfer feed between seasons or pastures, to extend feed supply and to balance pastures for the betterment of daily gains. This latter use is not generally considered in North America and it is the primary purpose of this publication. Unlike haylage and silage, dry hay can be fed on green pasture year around. Quality (TDN digestibility, protein) is the main factor that restricts or conditions hay use with finishing cattle. The greater the quality, the better the match with high-quality finishing pastures and the greater the response in weight gains. While haylage and silage can allow the making of stored forage in wet weather, a quality hay still should be your first choice with finishing cattle.

In most instances finishing quality hay will be legume hay, a mixed perennial grass and legume hay or hay made from cool-season annuals such as wheat, oats, cereal rye or annual ryegrass. Legume hays can be alfalfa, soybean, cowpea or clover based. I personally do not make any of the hay on my farm and use contractors exclusively. In North America you have the option of buying protein and digestibility tested (high relative feed value), dairy or finishing quality Western hay. If you live in a humid climate, buying in your finishing hay probably should be your first choice.

It has been estimated that in the Northeastern states, a forage producer only has about a 20 percent chance of making finishing quality hay. This is probably even lower if you plan to use contractors. Consequently, I do not plan to take up a lot of time with the mechanics of making hay, haylage or silage, as there are lots of references and extension resources for this. We want to primarily concentrate on pasture supplementation and its effect on raising average daily gain.

My stored forage goal is to have one 1200-pound

15

bale of high quality hay for each animal I am going to finish. This hay will be fed pretty much throughout the year in small amounts. Our purpose should not be to replace pasture but to supplement and complement it. There are weather emergencies when animals have to be finished exclusively on stored forages. Hopefully, such times will be infrequent and short as they are very expensive and do not produce the great quality meat.

Feeding Hay

Hay today is most commonly fed as round bales in hay circular feeders (grills) on pasture or in alleys. If paddocks are small, it is not uncommon to see hay provided in alleys so it does not have to be moved with each paddock shift. When feeding overnight in an enclosure, hay could be offered in a round-bale feeder or grazed from a hay stack (a line of round bales) with an electric fence (see diagram below).

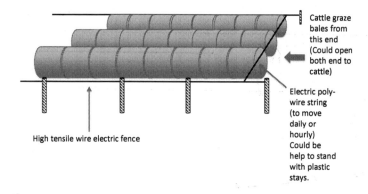

Cattle graze bales from this end (Could open both end to cattle)

Electric poly-wire string (to move daily or hourly) Could be help to stand with plastic stays.

High tensile wire electric fence

Line of round hay bales, rationed with electric fence.

This has the advantage of needing no tractor and hayfork lift or other large tools. The hay is located and piled during harvesting (at baling) where it will be utilized, and then the rate of feeding is managed with an electric fence structure. Cattle can have permanent or temporary access to the haystacks, similar to silage stack self-feeding systems seen with dairies.

Feeding hay under electric fence.

Unrolling hay for slice grazing with electric fence.

Alternatively, some people find it useful to un-roll the bales on the ground and slice graze the hay with an electric fence on top so that the animals do not trample, walk and lay

Examples of light weight hay rings.

on the hay. This is used also to patch/amend low fertility soil spots in the pasture with organic matter.

Since hay grills or feeders are the common tools to reduce hay feeding waste, make sure we pick a feeder that is simple, light to move, and put on the bale by hand, but still works.

Rings with round bale hay inserted.

Chopping hay can double the animal's voluntary intake

Hay quality implies protein and highly digestible fiber, which equates in high TDN. High relative feed value hay (or quality hay) is easily ruminated, which means the animal can break it down into small particles fast, which allows for more intake. As hay particle size diminishes, rumen contents mix faster which also helps to balance the diet if it is unbalanced in dry matter and fiber (e.g. nitrogen-rich washy pasture).

On the contrary, hay that is difficult to break into small particles has little chance for mixing. It floats on the rumen content. A floating mat of hay stems builds inside the rumen. As it struggles to ruminate it, the animal remains satiated and no diet balancing takes place. Sometimes, a compaction syndrome develops. Animals look full and the rate of fermentation in the rumen is slow. The rumen is full of indigestible forage and water. In the extreme, the animal weakens and could die of starvation, with its belly full.

Hay chopping could be contractor hired but most common is to have the one-bale-at-the-time chopper, which is run on the tractor power shaft.

Chopping the hay can yield a better use of it, prevent the compaction effect and increase intake. Chopped hay mixes better in the rumen, and does not trigger satiety as long hay does. Chopped to about one to two inches in length, it would not lose the effective fiber effect on rumen motility and rumination. Although all processing (chopping) increases utilization, high TDN hay would not require chopping (since the animal and the rumen bugs can easily reduce its size), but poor quality hay greatly benefits from chopping.

When chopping however, we need to be aware that some hays (namely legume hay) will develop dust and very fine particles (mostly from leaves). The chopping of such hays may reduce the quality if the leaves are lost into dust. Therefore, the speed and dryness of the hay should be looked at. A very dry and windy day can blow away lots of highly digestible fractions. Chopping legume hay does not improve intake as chopping grass hay does, and dust build-up is less relevant with grass hay.

In turn, chopping warm-season grass hay produces a *Hawaii* greater benefit compared with chopping cool-season grass hay. However, it has been noted that chopping triticale and oats hay can double voluntary intake of hay on pasture. Such effect has been reported to improve performance on winter annuals first grazing, improve transition between pastures of different quality, primarily going from warm-season to cool-season annual or perennial pastures.

Chopped hay has the disadvantage of the infrastructure required to store and feed. Feeding in a trough or a feed wagon is an extra cost and for many operations not worth the investment. However, figures have shown up to 25 percent improvement in hay use efficiency when chopped (higher intake and lower wastage). Chopping also reduces animal selection.

Simple distribution systems such as feeding below and behind an electric fence helps us feed without investing in several wagons or feeders. Limited amounts can be provided in long strips on a daily or every-other-day basis on pasture. The same hay can be used as a carrier for an energy supplement (solid or liquid).

We must keep in mind however, that chopping will not improve the quality or remove the constraints of moldy and spoiled hay. Chopping this type of hay should be avoided. Feeding moldy or spoiled hays should give the animals the opportunity of selection and avoidance of the worst fractions. In this case, chopping will reduce intake and increase the risk of intoxication.

A primer in the seasonal use of silage and hay in grass finishing programs

Finishing pastures are considered high quality pastures, which means, non-limiting crude protein, palatable, and highly digestible forage. All of these attributes result in a high daily intake of pasture. High quality pastures are well suited to Management-intensive Grazing without greatly affecting individual animal performance. But, as we pressure the animals to harvest more of the pasture sward, we make it harder for our animals to accomplish a high intake. Therefore, it is up to our management skills to allocate forage efficiently while affecting intake the least. Some seasons are more difficult than others for providing quality and quantity. Hay and silage are then used to provide volume for overcoming pasture shortage.

HAWAii DRought MGT!

We must remember however that animal performance on pasture with supplemental hay or silage is greatly determined by the positive association of both resources. In other words, a small amount of fresh pasture on a predominately hay or silage based diet will not improve high gains (2 lb/day gains), but it will improve low gains (1 lb/day gains). Before making our mind up about silage or hay and the type of each, it would be wise to analyze first the nutritional scenarios we will be dealing with. Although generalizations could be misleading, let's discuss probable seasonal scenarios.

SUMMER

In regions with hot summers, high rates of gain are very difficult on perennial pastures. Most people assume this is because the heat is making the animals reduce grazing hours, but it is also and maybe more important the heat's effect of the forage that is causing the low gains. Above 86 degrees F, perennial pastures (legume and grasses based) become less digestible and more lignified. Animals' weight gain will

HAY TESTS CONfiRM THIs!

respond to high quality supplemental forages. At very high temperatures (100 degrees F) all grass growth (even tropical grasses) halts. Grass quality becomes a relevant factor to determine which supplement is best. The supplement of choice to maintain adequate finishing gains during such hot weather *LEGUMES* periods would likely be alfalfa hay or pasture silage.

Alfalfa hay could be fed at a level of one third to one half of the daily intake (likely 1 to 2 percent of animal body weight on dry matter basis). A limited amount (less than 1 percent of the animal body weight) would not promote a significant change in daily gain. Our goal here is to improve or sustain average daily gains at 1.7 lbs per day as this is the minimum level to start laying intramuscular marbling fat. *ADG*

GAINS SUPREME

If alfalfa is the hay, positive interactions will occur and gains may improve because alfalfa becomes a protein and mineral supplement. Unfortunately, mixed grass/legume pasture hay will not have enough quality to meet this target. If hay is the only stored forage resource available, then we should feed the best quality hay we have. Low-quality hay will not help our animals to eat and digest more grass. If the hay is of poor quality, chopping it and adding a protein supplement *SOYBEAN* (soybean meal, DDGs) will help intake and digestion and it *CRYSTALYX IS + A APPROVED* will mix well with the grass. If you are opposed to the use of stored forages, you can consider using warm-season annuals such as no-grain, sterile corn, Sudan grass, crabgrass, soybeans forage as supplements, or timing your finishing period to cooler times of the year such as late spring or winter in the South of North America.

Operations in humid environments have difficulty making quality alfalfa hay and producers may prefer silage or haylage. Properly made, silages retain more nutritional value of the original plant compared to hay. Leaf loss is minimal for silage but significant for hay. On the other hand, if ensiling does not take place fast (pH does not drop), nutrient losses to respiration and undesired fermentations (butyric) could yield poorer feed than hay. It is important that acidity in the silage increase rapidly to slow sugar and protein degradation (pH needs to drop close to 4.0).

Cutting grass for silage.

Protein degradation results in less crude protein at feeding time, and also in a buildup of toxic substances such us nitrates, nitrites, ammonia and nitrogen oxides. If the amount of sugars in the silage mass is limited and pH does not drop below 5, checking for nitrate would be advisable. Most laboratories will run this test quickly and inexpensively.

Several silage gases are stringent and affect odor and palatability (sour silage), which affects intake, and some gases are toxic. Silage gas (nitrogen dioxide) has been known to kill cattle. The gas is heavy and highly toxic to humans and animals when present in concentrations greater than 10 to 25 ppm. In combination with water it causes pulmonary edema (pneumonia-like symptoms) and death.

Haylage is an intermediate option between hay and silage. It is harvested with 40 to 60 percent moisture. It is a drier feed than silage and retains more quality than hay. It requires bagging, and air infiltration is minimized. Due to the heavy use of plastic wrapping and the specialized equipment needed to make and move heavy haylage bales, haylage is the

Consolidating the silage stack.

most expensive stored forage. Its use must be limited to high value activities such as finishing.

Direct cut silage is another option, but alfalfa silage is more difficult to produce because of limited sugar content. Alfalfa is always short of sugar (5 to 9 percent of DM basis). Making good quality silage requires a minimum amount of readily available carbohydrates. Addition of at least 5 percent (DM basis) of molasses to the forage in layers (or sprayed) would help to make good alfalfa silage.

Most alfalfa and pasture silage is high in moisture content. Good alfalfa silage should start with dry matter content between 30 to 40 percent. This implies wilting prior to chopping if conventional silage is to be made from a lush alfalfa (more than 80 percent water content). Removing the air and maintaining an air-free atmosphere is more difficult in drier silage and it becomes a problem when dry matter is greater than 40 percent at ensiling. Thin chopping (implies better compaction at tractor packing) or vacuum pumping help improve the ensiling process.

Vacuum pumping the silage stack.

Silage with more than 65 percent water at feeding would be good to feed on a summer-dry forage diet, but we should keep in mind that more that 75 percent water in the silage almost always yields lower quality and reduces intake. Lower moisture silage results in better feed and improved animal performance.

In places where clover can be grown, white clover, red clover, ryegrass and brome mixed pastures can also make good quality silage suited for a finishing program. Legume and grass mixtures are easier to ensile than straight alfalfa as cool-season grasses are likely to have more than 10 percent soluble sugars. This forage needs to be cut earlier than pure legume stands, looking at the quality of the dominant grass in the mixture to determine cutting time. Dry matter is also likely to be higher in this silage, and it has a greater chance to stand alone as the only source of forage if pasture becomes scarce or inaccessible for a period of days.

Alternatively, silage from winter annuals is of finishing quality (up to 25 percent soluble sugars reported for silages from winter annuals such as annual ryegrass, cereal oats, triticale, barley and cereal wheat in some studies). Winter annuals are not often selected for silage production because of their low biomass production in a vegetative stage compared to leguminous pastures or summer forages (corn, sorghums, millets). The quality however, could be greater and can make good combinations with alfalfa hay or alfalfa silage. For best results as a stand alone source of feed, these silages should be produced during the vegetative stage (leafy stage) if dry matter content allows. You will read more about the need for adequate dry matter in forages for high levels of animal performance in the next segment.

FALL

Because grazing managers have traditionally sold calves and yearlings in the fall to reduce their winter stocking rate, they assume this would also be the optimum time to sell finished cattle as well. Actually, in many environments fall is the most difficult season to finish cattle. Places with a bimodal distribution of rainfall (primarily spring and fall) tend to have "washy" lush forage in the fall. This is not a problem on native range or warm-season pastures as it is on improved cool-season perennial legume and grass-based pastures, or on small-grain winter annuals. From a conventional laboratory analysis stand point; these forages look very high quality. They are high protein, low fiber, and highly digestible. Model predictions would indicate such forages should produce high intake and high gains. But, this is not always the case.

We often see depressed gains during this time of the year and great difficulty in finishing cattle. Very often the first frost " hardens" the forage. Dry matter content increases and fiber too, but most forages remain highly digestible and better balanced, and gains should improve. However, waiting for this effect to take place may not be practical if we have to graze animals on the pastures with no other option. Good quality hay plays an important role in improving fall gains.

If we are lucky enough to have fall rains, we may be able to avoid feeding hay during winter if snow depth allows the animals to reach the stockpiled grass. *Therefore, fall is the season in which feeding hay is probably most needed.* This is also true in the South where fall is the driest season. In this scenario, winter annual pastures growth is very slow depending on soil moisture, yet warm-season grass growth is slowing due to declining day length.

We can reduce animal stock to adjust forage supply to demand. We can even be conservative on demand and stock so as to have always forage available. But, adjusting the stocking rate is not effective at improving gains on washy pastures in the fall. Increasing stocking rates to eat more stubble or other residues would not work either to improve dry matter intake and gains, and it would be greatly detrimental to pasture persistence, soil cover and soil organic matter flux. On the contrary, residual biomass is needed as cover for soil and root protection from the subsequent freezing winter temperatures.

Feeding low quality hay does not help here either. Voluntary intake of poor quality hay is likely to be low and inconsistent. So, we need good quality hay for supplementing the "washy grass" syndrome. We can do well with pasture hay, which is less expensive and easier to make than alfalfa hay. Such hay should have enough quality for the animals to be willing to eat a quantity greater than one percent of body weight, while having lush forage the rest of the time.

For practicality, some producers prefer to provide hay in an enclosure. After the daily supply has been eaten, animals are released to pasture again. If this is the management we adopt, we may not want to keep the animals in the enclosure more than four hours. A longer period will affect the intake of pasture.

We must remember that lush pastures are mostly water (more than 85 percent) and animals have to harvest a large amount of forage to meet a high dry matter intake. As an example, an 800 lb steer, which would be willing to eat 24 pounds of dry matter a day, and eats one third in hay (8 pounds) needs to collect 16 pounds of dry matter while grazing. If the pasture has 18 percent dry matter, the animal will be harvesting 90 lbs of forage every day.

Palatable, high quality hays work best for this scenario. We need quality feed to match the quality the animals are grazing and compensate imbalances without depressing fiber digestion rate. Therefore, alfalfa hay, quality cool-season pasture hay, hay from small-grain winter annuals (hay from triticale, cereal oats, cereal wheat, annual ryegrass) would be preferred choice over hays from summer annuals (millet, sorghum, crabgrass, Johnson grass, bermudagrass, etc).

It is important that hay for this purpose is made from pastures in the vegetative stage. This is even more relevant for grass monocultures of annual species. Supplying daily an amount of hay or haylage between 0.5 percent up to 1 percent of the animal body weight (on dry matter basis) will be sufficient to counteract the high-moisture content and imbalance the nature of fall pastures.

In my personal experience, it has proven efficient to start the hay supplementation during the last days of summer before the onset of rains and washy feed. If we wait until we see loose feces to start supplementing, animal gains will have declined already. Supplemental feeding may be needed for three months in some areas with late frosts, and less than one month in other areas where frosts come early.

If it takes longer than three months and the amounts of hay fed are greater than one-third of the animal's daily intake, we may have to reevaluate the forage base and its economics. We may realize that the autumn of the year is not suited for finishing on pasture. While I personally sell finished cattle throughout the year, I skip the fall season due to its difficulty. Retreat is always a superior strategy to defeat.

During these lush-forage periods energy supplements such as sugar beet pulp, citrus pulp or molasses best combine with forages. A limited amount of readily degradable sugar (0.25 to 0.5 percent of body weight on a daily basis) could further improve gain without creating a disturbance in rumen pH, fiber digestion and meat characteristics.

Warning! While a little is good, a lot of energy supplement can be worse than none. Too much energy supplement may result in a temporary drop of rumen pH, retard fiber fermentation and affect the lipid profiles in meat or milk that make grass-fed products distinctive.

WINTER

Feeding massive amounts of stored forages in winter is commonplace even in America's subtropical regions where snow is rare. Surveys indicate that graziers feed hay to the same extent they can make it. Of course, it is more economical to stockpile perennial forages or plant winter annuals than to feed hay as the sole source of feed. Even in winter, stored forages are best used as a supplement to direct-grazed forages and not a full replacement of them.

Feeding stored forages on dormant native range would not be a sensible choice for finishing cattle. From a nutritional standpoint, feeding about half of the animal's daily intake in quality hay or silage will significantly improve gains. But, from a commercial standpoint, feeding a dry protein supplement would be more appropriate and will yield the same results, without the feeding expense of silage or hay.

Therefore, feeding hay or silage on winter-dormant pastures to improve gains makes little sense. If carrying capacity is the limiting factor (lack of forage), then the additive effect is needed and feeding the hay and silage would have better justification.

Dormant, stockpiled pastures are highly variable in quality. Native range species cure differently. Some retain quality and are well digested if a protein and mineral supplement is provided. The majority of cool-season species and some warm-season species adapted to high elevations and cold climates tend to cure better than the tropical ones found in the deep South of the USA. In most circumstances, however, these native range resources are far more appropriate for cow-calf and stocker programs than for finishing. An alternative approach would be to graze with a protein supplement and adequate stocking that avoids abusing the pasture. The rest of the animals could be fed in an enclosure on a hay or silage-based diet (or the combination of both).

In the case of alfalfa hay or silage, an energy complement would help to increase energy intake and improve gains, making a more efficient use of the stored forages. Feeding stockers on stored forages will not detrimentally affect the subsequent healthful meat characteristics we hope to achieve during the pasture-finishing phase.

Stocker programs can gain more than a pound a day on protein supplemented native ranges for several months, therefore, these ranges can provide a good winter base for subsequent finishing in spring and summer. However, my experience up to now indicates that finishing cattle on dormant native range is likely impossible in most environments.

Cereal rye, cereal oats, winter wheat, barley, annual ryegrass can all be grazed alone and require no supplemental feeding. Once the plants have passed the high-moisture stage early in the season, supplemental feeding of all kinds will only generate substitution and increase carrying capacity. Only during the first grazing at an early stage of growth of the winter annuals, supplemental feeding does improve animal performance. Hay and haylage are useful resources to balance the diet during that period. All the considerations about the lush grass syndrome discussed previously apply here.

Swathing and wilting for 24 to 36 hours, and grazing the windrows of swathed forage could be an additional tool to use to increase the dry matter and gains. This will also significantly help reduce the amount of hay to be provided during this period.

Again, the quality of this hay needs to be good enough

to match the quality of the pasture you are grazing. If quality of hay is poor, its fibrous fraction separates from the rumen mix and it does not get ruminated and digested at the same rate of the grass. It builds a matt in the upper rumen of slow turnover. Therefore, it does not balance the forage in fiber and actually decreases intake (physical satiety effects).

The quality of the winter annual pasture will dictate which hay we are able to use. If quality of pasture is high, we can supply medium quality hay (from cool-season and warm-season improved pasture, cool-season native pasture, or summer annuals). The better the quality of hay is, the greater the animal performance. But, we have to keep in mind that weight gains of cattle on hay or silage-supplemented winter annuals will rarely exceed the gains possible on non-supplemented annual pasture (provided availability is not limiting).

Quantity of hay or silage is very flexible during this time. If the decision to feed hay or silage during this time of the year has been made, most likely it is because winter annual pasture will be limiting. Therefore, we will be feeding an important amount of supplemental feed. Levels of 1 to 2 percent of the animal's body weight (on dry matter basis) are needed to generate a significant increment in carrying capacity and sustain a well balanced diet.

In the higher level, two-thirds of the daily diet is provided in the form of hay or silage and one third comes from the winter annual pasture. At this level of supplementation, performance will be dictated by the quality (TDN) of the feed. The winter annual will compensate for the nitrogen and some vitamins, but it will not be able to compensate for poor digestibility of hay. Therefore, as the amount of hay-to-feed increases, the quality of the hay has to also improve. If we cannot provide that quality, digestible dry matter intake will decrease and gains will decline below those necessary for finishing.

In a scheme with such a high amount of supplemental feeding, we may want to analyze the economics of alternatively pen-feeding some animals on the stored forage and feed none or much less hay to those left on pasture. This alternative simplifies transportation of hay or silage.

An important drawback is that the strategy removes the desirable distribution of feces over the pasture. Also, uncoupling the pasture from the hay or silage diet could create a need for a protein concentrate if the hay or silage is limited in protein. Protein deficiency is common in hay and silages made from warm-season species (from millet, meadow, sorghum Sudan, Johnson grass, bermudagrass, corn, etc). Warm-season hay is not only marginal in digestible energy, but also short in protein for most growing-finishing categories.

SPRING

Digestibility and protein of most key species of native ranges are high in the spring and finishing gains are possible. Even tropical forages such as bermudagrass can finish cattle in the late spring, and adding annual legumes to the sward can extend this into early June. Areas with late spring-summer rainfall patterns and moderate summer temperatures (high elevation or northern latitudes) have pastures that maintain a high-gain quality through most of the spring and summer. This normally allows for a low-cost finishing window of four to five months. Unfortunately, such areas are geographically rare in the United States.

Hay and silage could help regulate spring forage supply, however, adding these stored feeds will not improve average daily gains. Keeping free choice hay available will expand carrying capacity and help compensate for forage allocation errors. This latter benefit is particularly important for new grass finishers who may accidentally hurt their cattle's gains by taking the pasture too short before moving them.

The lush-pasture scenario described earlier for fall can be a spring scenario too. Similar to fall pastures, many operations are faced with fast and lush re-growth of native or improved pastures in spring. This forage is likely to be highly digestible, but excessive in water content, low in fiber, high in crude protein and low in soluble sugars. Animals will take time to adjust to this washy grass and will gain little to no weight for a period of time.

These delayed gains may not be so pronounced if the cattle come from another green standing forage they have been grazing such as small-grain winter annuals. The shift is greater and adjustment takes longer if animals come from a stored forage diet or dormant stockpiled pasture. Quality hay can be of great use in this transition period. The hay would not need to be legume based because the pasture will have sufficient crude protein, but it should be fair quality digestible hay. This hay-feeding period in early spring probably will not be long. After 15 to 20 days, most pastures should have enough fiber and dry matter to become balanced, and dry forage (such as hay) may no longer be needed. A quantity of about 0.5 percent of the animal's body weight should be sufficient, but if grass is still short, more hay could be fed.

An alternative to feeding hay during this period is to complementary graze on a partial day basis on stockpiled winter annuals such as winter cereal rye, winter cereal oats or barley, Transitioning from 100 percent stored feeds, or dormant pastures into green growing pasture, will allow the use of winter annuals as a link for a smoother change (progressive adaptation to a higher nitrogen and water content).

As pointed out earlier, periods of pasture lushness are the time for providing complementary energy supplements such as molasses. However (although research is limited on this matter), we want to be cautious regarding the amount of energy feeding we do on pasture in spring. A limited amount of energy supplement will balance nutrient supply and improve weight gain, but we need to be very careful to not impact the ruminal fermentation of forage because the entire growing season is ahead of us.

Drylot Finishing

It may be due to weather extremes of drought or heavy snow that we may want to finish cattle on stored forages only.

Full feeding on hay or silage for high gains is a real challenge with finishing cattle. We need to know about the quality of the material we started with and fully assess the possibilities for good gains on our stored forages. Hay or silage will not be better than its original material, and will likely be a lot worse.

Under no-grain scenarios, legumes and cool-season grass mixes make better silages than warm-season species. Silages tend to be better than hays in general. But, in particular, we need to look closely at the quality of silage we produce. Only a few stored forages can be used as a single diet and sustain gains above a pound a day. High quality alfalfa, clover and ryegrass, cereal rye, cereal oats or annual ryegrass hay and haylage may be energy and protein balanced and rich in TDN to meet requirements for high performance. But, producing this high quality hay requires the proper pasture, timing and weather.

Weather is the conditioning factor that we can control the least and it is what makes hay quality so inconsistent. This is particularly true in humid climates. Therefore, haylage and silage are good alternatives that allow us to work around weather problems and expand the possibilities for storing quality forages. However, dry hay should always be our first choice with finishing cattle.

Voluntary silage intake is lower than hay intake of similar quality. Water content and acidity seem to be partly responsible for this effect. Therefore, if both exist, a mixture in similar proportions of hay and silage (on a dry matter basis) would help balance intake and quality. Chopped hay mixed with the silage will improve intake. However, feeding separated fractions is a viable alternative, for example, silage in the morning and hay on the afternoon.

Alfalfa or leguminous pasture hay combines well with grass silage. Alfalfa silage mixes well with millet, sorghum, bermudagrass or crabgrass hay. These warm-season forage hays are not stand-alone hay because they are too marginal in digestible energy (TDN) and too short in protein for most stocker and finishing animals.

Alfalfa hay and annual ryegrass silage make an excellent mixture with very high gain possibilities due to the high digestibility of both. Similarly, while no-grain, green leaf corn silage cannot stand alone because of its limited protein content, combined with alfalfa hay it makes a good finishing feed. Corn silage and alfalfa hay is the primary finishing feed in much of the Mediterranean region of Europe.

Silage makes a good carrier for additions such as vitamin, mineral, protein and energy supplements in dry form. The wet state of the mixture prevents separation of fractions in the mixer. Most forage silages are short of soluble sugars and combine well with supplemental high sugar sources such as molasses and sugar beet byproducts. Again, we want to keep the amount of the supplemental energy source at a low level to avoid negative effects on fiber degradation and change rumen fermentation patterns. Providing a quantity of supplemental sugar no greater that 15 percent would be on the safe side.

A drawback to the feeding of stored forages over 60 days is a decline in the healthy fat lipids that are widely advertised as a benefit of grass-finished cattle. These healthy fat lipids only accrue from green living pasture. Long-term drylot finishing on stored forages produces a very different kind of beef than that produced by the direct-grazing of green forages. This is not a problem with stocker cattle that will be subsequently finished on green living forage.

A few observations about stored forages and finishing cattle:

1. Quality should be your production priority over quantity.

2. Quality hay has at least 60 percent TDN.

3. Quality silage has at least 65 percent TDN.

4. Silage intake will not be limited by water content when dry matter exceeds 30 percent.

5. Silage with more than 75 percent water could worsen a "washy forage" scenario.

6. Legumes and cool-season grasses make better hay than warm-season summer grasses.

7. Silage from mixed legume and cool-season grass

pastures is easier to make than straight legume silage.

8. Brown mid-rib sorghums and green leaf corn make good silage. Complemented with a protein supplement such as alfalfa hay, finishing gains are achievable.

9. Winter annuals (cereal oats, triticale, wheat and annual ryegrass) can make excellent silage and hay.

10. Combinations of hay and silage could help supply a more complete feed.

11. Stocker programs can be based on stored forages with little exposure to pasture with no ill effect on subsequent healthy fat lipids.

12. Finishing programs could make use of limited amounts of stored feeds if strategically allocated on pasture.

13. Stored forages can help finishing programs to transition better in the forage sequence.

14. Stored forages can help to start high gains earlier in the season.

15. Stored forages can help to expand the growing season over a month or more (depending on previous gains and quality of the stored forage).

16. High-quality alfalfa hay, legume and cool-season grass silage and small-grain winter annual hay or silage can make stand alone feed.

17. Alfalfa silage complements summer pasture well.

18. Alfalfa or leguminous hay complements fall, early spring pastures, and "washy" forages.

19. Pasture hay and silage is better for winter feeding.

20. Energy supplementation is most valuable during periods of washy forage.

See the section on Temporary Feeding Enclosures for details and illustrations.

Feeding during long periods of rain and mud

There are often periods of extended rains in much of North America and yet few people consider what this is doing to their cattle's health and average daily gains.

Mud and wind are negative factors for animal health and performance. A drop in intake for a day produces little harm on gains but a longer period affects performance and subsequent meat quality. Intake deprivation for several days also affects the immune system and exposes our animals to respiratory and foot diseases.

Grazing on rainy days, or in mud, is not easy for our cattle. They drop intake and resume it after the rain is over or until they become adapted. Adaptation to grazing or feeding in the rain takes three to four days, and depends on the nature of the feed. Lush grass and grass silage are not as desired by cattle as a dry feed during long rain periods.

Dry hay is the most preferred supplement to wet pasture. Moisture in the air and the rain predispose the animal to crave dry feedstuffs. Since intake is likely to drop (to half), the energy content of the feed is very relevant. Hay needs to have both quality and palatability to promote intake. It has to be easy to eat so that animals do not have to stay at the feed bunk long. Small volumes of quality hay should generate enough energy for maintenance and growth.

When feeding processed feeds as supplement on pasture, it is important to provide dry feed daily, otherwise the wet feed will not be eaten and animals will go hungry with the bunks full of feed. A drained spot (high area) of the paddock or enclosure allows the animal to feed at ease. However, we must keep in mind that animals will look for such high areas to lay down. So we may want to feed on gentle slopes so that animals do not crowd on top of the hills or close by the feeding area.

Sheds and windbreaks are also useful.

Summer annuals such as forage sorghum or corn, early planted, are good alternatives to feeding hay during rainy days. These grasses are highly palatable (sugars) and palatability promotes grazing and voluntary intake. For example, a no-grain, grazing-corn paddock could be the option for finishers, while other classes remain on other pastures supplemented with hay. Without removing the animals from pasture for a long period, we may have to push them to eat the dry hay for half a day and adapt them to come to it.

A self-feeding dry-feed wagon would remove the constraint of feeding daily in the mud, but the wagon needs to be checked frequently for wet feed and ponding around it. Chopped hay with molasses and other energy and fiber mixtures are suggested as excellent feeds during wet weather.

Feeds offered in bunks need to provide energy but fiber also needs to be there to maintain rumination and fiber digestion function in the rumen. We do not want to adapt the animals' rumens to a high concentrate/ low fiber diet since they will have to continue grazing after the rainy period is over.

Additionally, feeding or supplementing homogeneous

Moveable hay feeding wagon.

groups is important. Self-feeders or conventional bunks are always in short supply for the young or timid animals. Out on pasture, competition is minimized and dominants may not be as aggressive when competing for feed.

Weaned calves, in particular, have a difficult time adapting quickly to new feeding schemes. Creep feeding (feeding supplements) calves on their mothers is a great tool to teach them to bunk feed or to use hay. Recent research suggests that early imprinting of habits is a great opportunity to reduce learning time later on.

An alternative to feeding hay is a protected high spot of pasture. This works nicely during persistent rain. Grassy paddocks with stockpiled grass should be kept in mind as a safety zone during long periods of rain. Lastly, vaccinations and treatments should avoid rainy periods since the immune system is likely compromised.

Ensiling sweet sorghums

The lack of energy in cool-season plants is a major production problem for both dairy and beef producers who are producing a certified no-grain product. A forage for these producers to consider is sweet sorghum silage. Properly supplemented with protein, sweet sorghums (without grain) can generate greater than 2 lbs per day on finishing cattle, and up to 3 lbs per day on stocker cattle. This is great news for parts of the world where fattening on pastures during winter is impossible, or there are periods of extreme droughts or hot temperatures.

Sweet sorghum, or syrup sorghum (*Sorghum bicolor L.*), is a fast growing warm-season grass that has the ability to store significant amounts of sugar in the stalks. On average, stalks can contain up to 75 percent juice, which has from 10 to 25 percent sugar.

Sweet sorghum has been used as animal feedstock in the past in the United States but was displaced by the grain producing species in the USA's grain-centric mindset. Following this displacement as livestock feed, sweet sorghums were relegated to the farmstead syrup industry. Interestingly, the syrup industry has generated improved varieties (high sugar yielding and highly adaptable). More recently, the search for bio fuel sources found the world of sweet sorghums. There is sufficient sugar in sweet sorghum to consider this species as an alternative to corn for the ethanol industry.

Sweet sorghum silage has a similar feeding value as corn silage, if varieties are well picked, but has no starch. It is less expensive than corn silage and adapts very well to dry-land farming or areas with limited irrigation (it needs 12 to 16 inches of rainfall during the growing season and requires only 50 lbs of nitrogen per acre. Sorghums are efficient nitrogen scavengers, exploring soils in depth. Consequently, soils do not need to be as rich and farming suited as for corn, allowing planting on marginal ground. As with most sorghum species the root system is aggressive and competitive, out-competing other species.

Sweet sorghums are grown extensively for syrup in the Southeastern United States including Texas, north to Wisconsin, and west to Kansas, Iowa and Minnesota. Kentucky and Tennessee are the leading states in syrup production.

Besides sugar, sweet sorghum contains nutrients as iron, calcium and potassium. Research in many countries has concluded that sweet sorghum silage can be of similar quality and result in similar animal gains as corn silage. The advantages over corn are that adaptability of sorghums is unparalleled by corn, it is a less expensive crop, and the harvesting window is more flexible since quality is not based on starch in grain and grain content.

A.) Given a properly planted and well-nourished crop, moisture content at harvesting for silage would be the number one concern. A common error at ensiling sugar sorghums is cutting with excessive plant moisture. Dry matter content should be above 30 percent (less than 70 percent water in the whole plant) and preferably closer to 35 percent. Otherwise, fermentation becomes too acidic (beyond lactic acid), water is not retained in the plant mass and gravity flows to the bottom of the pit (or bag) and drains away washing nutrients and lactic acid.

Silages with excessive water slow voluntary intake and depress weight gain. Palatability issues have also been a problem when high water silages are fed to light calves. The second limiting factor to animal performance is protein content.

Sweet or green-leaf sorghum silages are very low in crude protein. Corn or grain sorghum silages have low protein contents (7.5 to 8.5 percent on dry matter basis), but syrup sorghums are likely to have less than 5.5 percent and often 4.5 percent crude protein. This is a limitation that needs to be remedied to fully utilize the energy in the silage.

B.) Sweet sorghum silages are a good alternative to balance the diet and expand carrying capacity of winter annuals such as winter wheat, triticale, barley or ryegrass. Feeding schemes based on hourly grazing on winter annuals and self-feeding on sweet sorghum silage are an effective low input strategy implemented by producers in South America.

Conversely, bunk feeding of fixed amounts of sweet sorghum silage to animals grazing winter annuals is also a good alternative. In this case, silage has to be provided daily in an enclosure to encourage the animals to eat it. Likewise, sweet sorghum silage would be the silage of choice to improve performance on alfalfa pastures during a late summer slump.

C.) A third factor to consider is particle size. Chopping to a small particle size is needed. Being a highly fibrous material and intended as the main component of the diet, effective fiber will be in excess for the rumen requirements of beef cattle, even at the smallest particle size the commercial machinery can produce. Small particle size will allow for packing the silage well and preserving the plant sugars and proteins. Reducing particle size to meet the targeted dry matter content (mentioned above) is the challenge for contractors. As dry matter content increases, chopping to small particle size becomes more difficult.

Sweet sorghum silage combines well with high quality alfalfa hay, and all protein rich rumen degradable by-products, as protein sources. Calculating the protein content of the diet against the animal's requirement is central to performance on sorghum silage. Gains on protein uncorrected silage could be less than 0.3 lb per day, or weight loss.

The improved lines have the capacity of germination at lower soil temperatures than in the past (closer to 55 degrees F), which allows us to plant in April in the colder climates (Nebraska), basically at the same time as corn. Also, if sterile F1 hybrids and non-flowering types are adapted to the area, the chances of producing a silage with less lignin and a wider harvesting window increase. The brown midrib (bmr) mutants of sorghum have lower levels of lignin content (50 percent less in the stems and 25 percent less in the leaves). Variability of cultivars is large however and need to be locally identified.

Sweet sorghum harvested for silage can be harvested before plant maturity (after reaching the minimum dry matter content). Therefore, it is possible to grow it in a double-crop sequence with a winter annual (eg. triticale, cereal rye or annual legumes).

Hairy Vetch Can Provide The Nitrogen

Hairy vetch (*Vicia villosa Roth*) is a winter-hardy legume that can be seeded in fall and reach maturity by early summer. Hairy vetch is capable of producing as much as 3-5 tons/acre and leave up to 100 lbs of nitrogen per acre in the soil, which is sufficient to grow a full crop of sweet sorghum.

Vetch should be planted in September or October after the sweet sorghum harvest, and used for silage, balage or harvested in late June.

(Planting times differ greatly however with the region and rotation schemes.)

Hairy vetch with triticale or cereal rye (mixed) seeded in fall after the sweet sorghum crop, could make a good rotation to maximize forage yield and reduce nitrogen inputs. A barley and vetch mixture (planted in fall) could be an alternative also to generate a barley plant and vetch silage for summer use while the sweet sorghum would be the silage for winter.

Planting sweet sorghum with peas (peas every third row) or over seeding peas after planting the sorghum has been researched. Either option was promising to increase crude protein in the silage to generate a "complete feed." Timing the species for cutting at the adequate maturity stage is key, and defines of quality. The combination of the two sources (sorghum and peas or soybeans) could be produced at the time for chopping and ensiling. Two separate fields are easier to manage than a mixed one. Gaining experience on producing straight sweet sorghum silage first, before trying the mixtures, would be recommended.

Because one does not have to target for grain content, the planting period becomes more flexible than for grain-rich sorghum or corn silage. Sweet sorghum could be planted in most regions from late April to the end of May or even early June. Most sweet sorghums would be ready to chop for silage within 80 to 100 days after germination (variety and region depending).

Keep in mind that the later we plant, the lower the biomass yield is likely to be. But, for practical purposes a forage harvesting window could expand as far as a week after the first frost. Sorghums are highly sensitive to frost. Frost stops plant respiration and will kill the plant. Disintegration of active tissue initiates quickly and dehydration starts in minutes, but degradation of nutritional value happens more gradually and has been found to be of little significance during the first five days after frost, which gives us time to produce silage.

More than nutrient loss, plant dehydration would be our concern in the demand for rapid ensiling before dry matter content becomes excessive. Conversely, a frost could be used as a strategy to dry the crop to a targeted dry matter content.

Temporary feeding pens can build soil while feeding cattle

Electric fenced, temporary feeding enclosures are a great tool to feed supplemental forages inexpensively while simultaneously improving the organic matter and nutrients of poor soil areas on your farm or ranch. Rather than gathering and moving the manure as is done with permanent feedlots, the feeding area is moved and the manured area is planted to high-nutrient-demand crops or forages.

Feeding stored forages on a given section of land can increase soil organic matter far beyond what we can do over a much longer period with conventional grazing. Silage or high hay diets are the preferred feeds since they generate larger quantities of manure and are of greater carbon content (fiber) than high-grain ones. Increased deposition of fiber increases water retention, soil air flow, and counteracts soil compaction. Active carbon sources in the soil and access to air favors soil micro-flora development to capture nutrients and build soil organic matter.

Experiences with farms of central Argentina (climatically similar to central Oklahoma) have shown significant improvements in fertility and soil organic matter after one or two years of winter feeding on sandy soils. After the feeding period, crop yields (soybeans and corn) on the sandy slopes matched yields from the farms' best soils. No contamination (nitrate enrichment) of the water table was detected. Crops were healthy and no extra fertilizer was needed for five years. Organic matter increased three times and phosphorus (low P soils) 10 times. In contrast, areas heavily "manured" for two years showed sectors of excessive nitrogen and slight soil acidity. On these areas crops reached peak production on the second year of cropping.

No over-enrichment and no deleterious effects on the first crop were found if areas were used for feeding for only one year. After the feedlot manuring, conservative cropping, including use of winter cover crops, was central to maintaining the improved soil structure and production from such poor soils.

Here are a few tips that we have learned over the years:

A.) Feeding pens are planned for only one or two seasons (one or two years of use) and then moved to another area. Since no feedlot floor is built, manure and urine will mix with the soil. Consequently, a longer use period will tend to produce soil over-enrichment and over-compaction. Unlike permanent yards, temporary penning areas will be cropped or pastured, therefore excessive manure buildup is not a risk

B.) The water table is the first concern. A minimum depth to the water table of five feet is necessary to control contamination of the underground water. If soils are shallower and sandy, moving the location of pens annually is recommended.

C.) The chosen area should have slope to avoid water ponding and permanent mud. Infiltration of nutrients will take place and fast leaching creates the risk of water table contamination. This is likely to occur if the water table is too close to the surface and pens are used for several years. Slopes should be gentle (0.25 to 0.8%) but are needed to keep the pens dry and prevent ponding. Since the soil is to be cropped afterward, the objective is to build soil structure and fertility, soil moving and mound building is not an option. Accumulations of manure near fences, water and feeding points should be re-distributed in the pen, but not removed unless excessive.

D.) A low animal density (about half of a regular earthen permanent feedlot) is recommended. However, the correct animal density depends on soil type, slopes and rainfall. Sandy soils in rainy environments should be allowed lower densities compared with clay soils and arid environments. Land allocation would range between 250 to 200 square feet per animal.

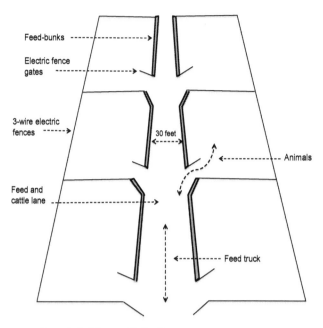

Feed-bunks

Electric fence
gates

3-wire electric
fences

30 feet

Animals

Feed and
cattle lane

Feed truck

Figure 1. Building mobile feeding pens

E.) Pen size is also important. Pens larger than 150 x 1560 feet are not recommended. Temporary electric fence pens should hold less than 100 animals. Planned mobile infrastructure is likely to be fragile and inexpensive, therefore small numbers and easy handling is a key element. No dogs, gentle routines and few people are fundamental to success with minimalist pens.

F.) Farm experience has shown that the entire infrastructure can be built out of electric fence. A 3-wire fence works well. Posts with insulators should be placed every 30 to 50 feet and three wires are enough to separate pens. The quality and strength of wire is critical and high-tensile wire is recommended. Wire strength is needed to withstand weathering and to hold up the birds that will perch on the wire. The wires have to withstand strong winds and the sudden strike of frightened animals. Although stressed animals could easily get through, it is important that the wire does not break. Width between wires on the fence depends on cattle size.

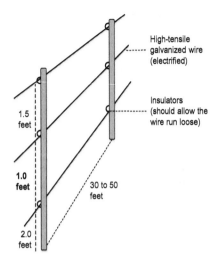

Figure 2. Building temporary fences for enclosures or feed-lots

G.) Lanes to feed and move the cattle can be the same ones since pens will be moved. Feeding lanes are not permanent. Moving cattle through a central lane is safer than on the outside since no permanent fences surround the structure. An outside continuous fence could be built but unless sensitive areas are exposed (roads, houses, high-value horticultural crops), most operations do not find the need. Animals that get out are easily walked back via the center lane to their respective pen. Angled gates help to create an easier walkway into the pen.

H.) Feed-bunks are also temporary and do not need to be continuous along a feeding lane. Feed-bunks could be built out of most materials, wood, aluminum, plastic drums, etc., but functionality is the key. Since bunks are to be removed, they need to be light, easy to pick up, to transport and to store. And more importantly, they need to be inexpensive. We must keep in mind that these yards are built for opportunity use such as the spring, winter feeding or during severe drought and not

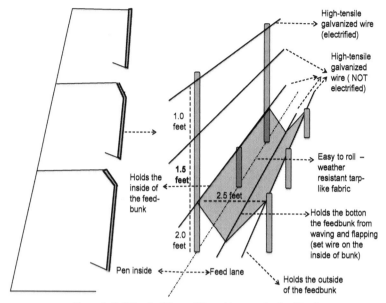

High-tensile galvanized wire (electrified)

High-tensile galvanized wire (NOT electrified)

1.0 feet

1.5 feet

2.5 feet

2.0 feet

Holds the inside of the feed-bunk

Pen inside ← → Feed lane

Easy to roll – weather resistant tarp-like fabric

Holds the botton the feedbunk from waving and flapping (set wire on the inside of bunk)

Holds the outside of the feedbunk

Figure 3. Building flexible, mobile and inexpensive feed-bunks

for continuous use. Often, temporary feed-bunks in Argentina are built out of strong tarp-like fabric. Most weather resistant fabrics work well. The material needs to be strong enough to hold feed and last intact for at least one year. It should also be easy to fix if broken (by the feed truck, animals stepping, or the wind). Performance is better if the edge wires can be threaded the length of the piece. Sections longer than 80 feet are difficult to deal with. Each feed-bunk will need more than one piece, and at least two if of 50 feet long. A wire extended the full length of the bunk at the bottom on the inside of the bunk will restrict flapping in the wind, which is the main cause of rupture.

I.) Water distribution is often a limiting factor. If water pipes are nearby, then water troughs can be temporarily set in each pen without an additional reservoir. In most situations, however, a large reservoir needs to be placed on each side of the pen lines (on the opposite side of the main lane) and serve the drinking troughs. Keeping this tank full requires planning.

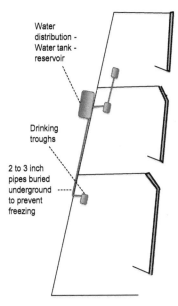

Water
distribution -
Water tank -
reservoir

Drinking
troughs

2 to 3 inch
pipes buried
underground
to prevent
freezing

Figure 4. Building temporary water supply for enclosures or feed-lots

Commonly eight to ten gallons of water per animals per day are used as a guideline for the water infrastructure.

J.) Creating a living green "filter" of grass downslope would be ideal. Additionally, capturing run-off water (after a rain) gives the opportunity of distributing the manure over a larger area beyond the area with pens. Run-off points from the pens need to be detected and mapped. A land survey should provide us the lowest and highest points and help us to develop a gravity-driven, irrigation-like run-off distribution system to distribute rainwater as evenly as possible on the area surrounding the feedlot.

An area twice as wide as the pens has proven to capture most of the visible runoff. Row crops (corn and soybeans) planted under no-till practices and after a winter cover crop (cereal rye) on these green filters have matched top yields of the same property. It would be recommended to use to fast growing, high biomass, nutrient demanding forage species, which preferably should be harvested for silage or hay. This would help to capture and remove the greatest amount of nutrients possible while leaving a large root system.

K.) Lastly, monitoring the water table for nitrates would be a routine to perform twice a year. Nitrate enrichment of the water table is the main concern when pen feeding and another indicator that may tell us we should move to a different sector of the property.

Index

About the author

Dr. Anibal Pordomingo graduated from the National University of La Pampa in Argentina with a degree in Agronomy Engineering. He received a Master's Degree in Animal Nutrition from New Mexico State University and doctorate in Animal Science.

In addition to teaching as an Associate Professor at the School of Veterinary Sciences at the University of La Pampa, he is the Senior Researcher at the National Institute of Agriculture Research of Argentina (INTA). His research covers animal nutrition, production systems, feeds and feeding, and beef quality on grazing. Speaking engagements have taken him throughout South and North America, to France and Scotland. He has published numerous scientific papers and articles and is a frequent contributor and presenter for the *Stockman Grass Farmer.*

With his family he runs two farms of 1200 acres total with 260 cows in a cow-calf and stocker finishing program. 240 steers and heifers are grown and pasture finished each year in a 27-inch annual rainfall area using an improved perennial and annual pasture forage base.

He lives in Santa Rosa, Argentina, with his wife, Elisa Nilda Arese, and son Martin Gabriel Pordomingo.

Questions
about grazing ???????
Answers *Free!*

While supplies last, you can receive a Sample issue designed to answer many of your questions. Topics include:

* How You Can Beat the High Cost of Cow Depreciation
* What Is Your Livestock Business' Breeding Objective?
* Selling Grassfed Beef on Flavor and Production Practices
* Three Proven Prototypes for Pastured Poultry
* Recycle Nutrients
* High Stock Density Grazing and Daily Shifts
* Underappreciated Weeds
* The Affects of Soil Acidity on Grazing Animals
* And more

Green Park Press books and the Stockman Grass Farmer magazine are devoted solely to the art and science of turning pastureland into profits through the use of animals as nature's harvesters. To order a free sample copy of the magazine or to purchase other Green Park Press titles:

P.O. Box 2300, Ridgeland, MS 39158-2300
1-800-748-9808/601-853-1861

Visit our website: www.stockmangrassfarmer.com
E-mail: sgfsample@aol.com

More from Green Park Press

AL'S OBS, 20 Questions & Their Answers by Allan Nation. By popular demand this collection of Al's Obs is presented in question format. Each chapter was selected for its timeless message. 218 pages. **$22.00***

COMEBACK FARMS, Rejuvenating soils, pastures and profits with livestock grazing management by Greg Judy. Takes up where *No Risk Ranching* left off. Expands the grazing concept on leased land with sheep, goats, and pigs in addition to cattle. Covers High Density Grazing, fencing gear and systems, grass-genetic cattle, developing parasite-resistant sheep. 280 pages. **$29.00***

GRASSFED TO FINISH, A production guide to Gourmet Grass-finished Beef by Allan Nation. How to create a forage chain of grasses and legumes to keep things growing year-around. A gourmet product is not only possible year around, but can be produced virtually everywhere in North America. 304 pages. **$33.00***

KICK THE HAY HABIT, A practical guide to year-around grazing by Jim Gerrish. How to eliminate hay - the most costly expense in your operation - no matter where you live in North America. 224 pages. **$27.00*** or Audio version - 6 CDs with charts & figures. **$43.00**

KNOWLEDGE RICH RANCHING by Allan Nation. In today's market knowledge separates the rich from the rest. It reveals the secrets of high profit grass farms and ranches, and explains family and business structures for today's and future generations. The first to cover business management principles of grass farming and ranching. Anyone who has profit as their goal will benefit from this book. 336 pages. **$32.00***

LAND, LIVESTOCK & LIFE, A grazier's guide to finance by Allan Nation. Shows how to separate land from a livestock business, make money on leased land by custom grazing, and how to create a quality lifestyle on the farm. 224 pages. **$25.00***

MANAGEMENT-INTENSIVE GRAZING, The Grass-roots of Grass Farming by Jim Gerrish. Details MiG grazing basics: why pastures should be divided into paddocks, how to tap into the power of stock density, extending the grazing season with annual forages and more. Chapter summaries include tips for putting each lesson to work. 320 pages. **$31.00***

More from Green Park Press

MARKETING GRASSFED PRODUCTS PROFITABLY by Carolyn Nation. From farmers' markets to farm stores and beyond, how to market grassfed meats and milk products successfully. Covers pricing, marketing plans, buyers' clubs, tips for working with men and women customers, and how to capitalize on public relations without investing in advertising. 368 pages. **$28.50**

NO RISK RANCHING, Custom Grazing on Leased Land by Greg Judy. Based on first-hand experience, Judy explains how by custom grazing on leased land he was able to pay for his entire farm and home loan within three years. 240 pages. **$28.00***

PADDOCK SHIFT, Revised Edition Drawn from Al's Obs, Changing Views on Grassland Farming by Allan Nation. A collection of timeless Al's Obs. 176 pages. **$20.00***

PASTURE PROFITS WITH STOCKER CATTLE by Allan Nation. Profiles Gordon Hazard, who accumulated and stocked a 3000-acre grass farm solely from retained stocker profits and no bank leverage. Nation backs his economic theories with real life budgets, including one showing investors how to double their money in a year by investing in stocker cattle. 192 pages **$24.95*** or Abridged audio 6 CDs. **$40.00**

QUALTIY PASTURE, How to create it, manage it, and profit from it by Allan Nation. No nonsense tips to boost profits with livestock. How to build pasture from the soil up. How to match pasture quality to livestock class and stocking rates for seasonal dairying, beef production, and multispecies grazing. Examples of real people making real profits. 288 pages. **$32.50***

THE MOVING FEAST, A cultural history of the heritage foods of Southeast Mississippi by Allan Nation. How using the organic techniques from 150 years ago for food crops, trees and livestock can be produced in the South today. 140 pages. **$20.00***

THE USE OF STORED FORAGES WITH STOCKER AND GRASS-FINISHED CATTLE. by Anibal Pordomingo. Helps determine when and how to feed stored forages. 58 pages. **$18.00***

* All books softcover. Prices do not include shipping & handling

To order call 1-800-748-9808
or visit www.stockmangrassfarmer.com

Name _____

Address _____

City _____

State/Province_____Zip/Postal Code _____

Phone _____

Quantity	Title	Price Each	Sub Total
____	**20 Questions** (weight 1 lb)	**$22.00**	_____
____	**Comeback Farms** (weight 1 lb)	**$29.00**	_____
____	**Grassfed to Finish** (weight 1 lb)	**$33.00**	_____
____	**Kick the Hay Habit** (weight 1 lb)	**$27.00**	_____
____	**Kick the Hay Habit Audio - 6 CDs**	**$43.00**	_____
____	**Knowledge Rich Ranching** (wt 1½ lb)	**$32.00**	_____
____	**Land, Livestock & Life** (weight 1 lb)	**$25.00**	_____
____	**Management-intensive Grazing** (wt 1 lb)	**$31.00**	_____
____	**Marketing Grassfed Products Profitably** (1½)	**$28.50**	_____
____	**No Risk Ranching** (weight 1 lb)	**$28.00**	_____
____	**Paddock Shift** (weight 1 lb)	**$20.00**	_____
____	**Pa$ture Profit$ with Stocker Cattle** (1 lb)	**$24.95**	_____
____	**Pa$ture Profit abridged Audio -- 6 CDs**	**$40.00**	_____
____	**Quality Pasture** (weight 1 lb)	**$32.50**	_____
____	**The Moving Feast** (weight 1 lb)	**$20.00**	_____
____	**The Use of Stored Forages (weight 1/2 lb)**	**$18.00**	_____
____	Free Sample Copy ***Stockman Grass Farmer*** magazine		_____

Sub Total _____

Mississippi residents add 7% Sales Tax _____

Postage & handling _____

TOTAL _____

Shipping	Amount
Under 2 lbs	$5.60
2-3 lbs	$7.00
3-4 lbs	$8.00
4-5 lbs	$9.60
5-6 lbs	$11.50
6-8 lbs	$15.25
8-10 lbs	$18.50

Canada
1 book $13.00
2 books $20.00
3 to 4 books $25.00

Foreign Postage:
Add 40% of order
**We ship 4 lbs per package
maximum outside USA.**

www.stockmangrassfarmer.com

Please make checks payable to

**Stockman Grass Farmer
PO Box 2300
Ridgeland, MS 39158-2300**

**1-800-748-9808
or 601-853-1861
FAX 601-853-8087**